The HOUGHTON M GUIDE to the INTEI for CHEMISTRY

James A. Dix
State University of New York--Binghamton

HOUGHTON MIFFLIN COMPANY   BOSTON   TORONTO
Geneva, Illinois   Palo Alto   Princeton, New Jersey

Editor-in-Chief: Kathi Prancan
Editorial Assistant: Stacey Pastor
Production Coordinator: Deborah Frydman
Senior Manufacturing Coordinator: Priscilla Bailey
Marketing Manager: Karen Natale

The on-line version of this guide is available at http://www.hmco.com/ and is updated periodically.

Copyright © 1996 by Houghton Mifflin Company. All rights reserved.

No part of this work may be reproduced or transmitted in any form or by any means, electronic or mechanical, including photocopying and recording, or by any information storage or retrieval system without the prior written permission of Houghton Mifflin Company unless such copying is expressly permitted by federal copyright law. Address inquiries to College Permissions, Houghton Mifflin Company, 222 Berkeley Street, Boston, MA 02116-3764.

Printed in the U.S.A.

ISBN: 0-395-78822-6

23456789-PR-00 99 98 97

CONTENTS

Introduction 1
What is the Internet? 1
Internet Tools 2
    World Wide Web 3
    Electronic Mail 3
    FTP 4
    Telnet 4
    Gopher 5
    Electronic Discussion Groups 5
    Audio and Video Conferencing 6
Connecting to the Internet 6
    Hardware 6
    Software 7
        Driver Software 7
        World Wide Web Browsers 8
        Browser Helper Programs 9
        Freeware and Shareware Programs 10
        Applications for Mac Platform 11
        Applications for PC Platform 11
        Commercial Software 12
    Internet Service Providers and On-line Services 12
Searching the Internet 13
    Finding People 13
    Finding Information 14
The Internet in Chemistry Education 15
    Course Material 15
    Electronic Discussion Groups 17
    On-line Journals 20
    Electronic Conferences 21
    Creating and Publishing World Wide Web Content 21
    General Interest 22
Advantages and Disadvantages of the Internet 22
    Security 22
    Reliability 23
    Information Overload 23
    Copyright Issues 23
The Future of the Internet 24

INTRODUCTION

The Internet today is fundamentally a communication medium. Its lineage can be traced back to the earliest forms of spoken and written communication, and through the development of mail, telegraph, telephone and television. The Internet bundles most methods of communication in one electronic package. With the Internet, people can use the communication media that have developed over centuries to conveniently interact from anywhere in the world.

The advantages of the Internet for chemists are many fold. For example, an Internet search from a desktop computer in December, 1995, of 3,300 scientific journals and 1×10^7 computer files for the key words "combinatorial chemistry" found 49 journal references and 16 computer files in less than ten seconds. Within minutes, a scientific result can be made available world-wide. Video from the space shuttle and photographs from the moons of Jupiter are available on the Internet almost as soon as they are received on earth.

The Internet also provides new modes of chemistry instruction. For example, an instructor can distribute laboratory instructions, including video of key procedures, via the Internet. The time-consuming steps of duplication and distribution of printed instructions are, thus, eliminated, as students have on-demand access to text, video, audio, and images pertaining to the laboratory. Course materials and lab instructions can be updated quickly and easily; instructors no longer need to provide printed updates to students. Other instructors anywhere in the world can review lab experiments and provide feedback on them. Students can post questions via e-mail to an electronic bulletin board and receive answers from instructors or other students world-wide. Instructors no longer need to hold office hours; they can answer all questions by posting messages to the bulletin board. Electronic mail, thus, frees all parties from having to be at a particular place at a particular time to communicate.

This guide is an introduction to how the Internet can be used in chemistry education. The Internet is new, and exciting, and it is changing rapidly. What a chemistry instructor can do with the Internet is limited only by his or her imagination.

WHAT IS THE INTERNET?

The Internet began in 1969 when the Advanced Research Projects Agency (ARPA) of the Department of Defense (DOD) decided to connect computer networks at military and defense institutions with those at universities that did contract work with the DOD. The initial goal of the project was to allow researchers at one institution to run programs on computers at another institution. Because computers made by one manufacturer could not communicate with those made by another, hardware and software standards were developed to transfer information between computers. These standards, collectively called the Transmission Control Protocol/Internet Protocol (TCP/IP), form the common language of computers connected to the Internet.

As computers became more commonplace and the advantages of connecting computers together became apparent, networks of networks became more widespread. ARPA established ARPANET in 1983 for non-military communications. The National Science Foundation created NSFNET in 1986 to connect six supercomputers around the country, forming a computer network backbone to which other computer networks soon became attached.

ARPANET and NSFNET were created to be used only for research. However, as the number of interconnected networks increased, the federal government recognized the potential of the Internet as a commercial distribution network. On May 1, 1995, the National Science Foundation turned over management of NSFNET to commercial organizations, ending 26 years of the federal government's primary supervision of a major portion of the Internet.

From a handful of computers, the Internet has grown to more than six million machines (**http://www.nw.com/zone/WWW/report.html**)[1] connected by a conglomeration of wires, optical fibers and wireless links. The connections are owned and operated by organizations such as Sprint (**http://www.sprintlink.net**) and NYSERNet (**http://www.nysernet.org/about/index.html**). These companies sell Internet access directly to large universities or businesses, or to resalers, who offer regional and local access. For users in academia, access to the Internet appears to be free, but it is not; money is transferred from the universities to Internet service providers.

While there is no formal structure to the Internet, there is an Internet Society (**http://info.isoc.org/**) that coordinates efforts among various researchers to develop Internet standards. These standards are important to maintain order. For example, each packet of information sent out on the Internet contains two unique numbers representing the address of the sender and the address of the receiver. If there were no central authority to assign a number to one and only one computer, a packet of information could not be identified as coming from a single computer nor would there be a guarantee that the information would be delivered to the intended receiver. An analogous situation would be if the same zip code were assigned to several different locations across the country. The registration of addresses is currently provided by Network Solutions Inc. (**http://rs.internic.net**) under contract to the federal government. While in the past registration was free, there is now a fee to register Internet addresses (**http://rs.internic.net/announcements/fee-policy.html**).

INTERNET TOOLS
The Internet provides a set of tools that can be used to enhance chemistry education. These tools are summarized briefly below. The section The Internet in Chemistry

---

[1] Note: In this guide, references to additional information are given in the form of Uniform Resource Locators, or URLs. The URLs typically start with characters such as http:// and ftp://, and are "addresses" to files on computers that can be accessed using a World Wide Web (WWW) browser such as Netscape Navigator or NCSA Mosaic, which are described in the section below. The URL can be typed into the browser to access the information. If you are reading the on-line version of this guide using a graphical WWW browser, you can access the information by clicking the mouse on the URL.

Education gives some specific ideas on how to use the tools in teaching and learning chemistry.

World Wide Web
The World Wide Web (WWW) (**http://www.w3.org/**) was started in 1989 at CERN (the European Laboratory for Particle Physics) (**http://www.cern.ch/**) as an easy way to use the Internet to share information stored on different computers.

Navigation within the WWW is through hypertext--text linked to other text and resources. Selection of a certain word in a hypertext document--by clicking on the word with a mouse or selecting it with keyboard commands--immediately links the user to another document or file. In this way, the user is able to access documents and files stored on computers across the world.

Although the concept of hypertext was not new in 1989, the WWW extended the concept to include documents stored on a network of computers, not just ones stored on a single computer.

The WWW would have remained just another way to access information from the Internet were it not for the killer application of the decade, NCSA Mosaic (http://www.ncsa.uiuc.edu/SDG/Software/Mosaic/NSCAMosaicHome.html). NCSA Mosaic provided an intuitive, graphical way to browse the Internet, and was as revolutionary to the Internet as the Macintosh operating system was to personal computers in 1984. With NCSA Mosaic, the Internet became accessible to a much wider range of people because users no longer needed to know the sometimes arcane language of the Internet. For example, to retrieve a file from the Internet in the pre-Mosaic days, a user had to know such things as the difference between ASCII and binary files, and commands such as lcd, ls, and get. NCSA Mosaic replaced these steps with a single mouse click.

Most information transferred over the Internet is transferred via the WWW; as a result, the Internet and the WWW have become nearly synonymous. Equipped with a program to browse the WWW (see the section World Wide Web Browsers, below), chemistry instructors can easily locate and access information that can be used in teaching chemistry. With some additional investment in time and money, chemists can also create and publish content on the Internet that can be accessed by students and colleagues world-wide (see the section Creating and Publishing World Wide Web Content below).

Electronic Mail
Before computer networks became widespread, there were three modes of interpersonal communication, in order of decreasing capacity to convey information. These modes were face-to-face interactions, telephone conversations, and mail. The Internet has advanced another method of communication: electronic mail. While lacking the real-time information exchange that occurs over the telephone, e-mail allows users to exchange many more messages in a given period of time than regular mail (sometimes called snail mail). E-mail also saves time in that the additional steps

of printing out a letter, enclosing it in an envelope, adding a stamp, and depositing the letter in a mailbox are eliminated.

To use e-mail, a user types a message on a computer terminal, then sends the message to an individual or a group of people. In more sophisticated e-mail applications, users can attach computer files, such as Microsoft Word documents or Lotus 1-2-3 spreadsheets, to the messages. The file is sent along with the message and can be retrieved if the recipient has the appropriate software application to view the file. E-mail is also used to participate in electronic discussion groups, as described in the section, Electronic Discussion Groups, below.

FTP

There is a myriad of computer files available to chemists, ranging from course outlines to molecular modeling programs to digital multimedia. Many of these files can be viewed using a World Wide Web Browser (see section below). Files can also be transferred (downloaded) from a remote computer to a local computer (the client computer) using a program called file transfer protocol (FTP). Users can also place (upload) files to a remote computer (the server computer) using FTP.

There are two different ways to access files via FTP. One way is by anonymous FTP, which enables anyone to access files by logging on to the server computer using the userid **anonymous**. The Freeware and Shareware Programs, described below, are available by anonymous FTP. The other way to access files requires a userid and password. These can be provided by the administrator of the computer on which the files are stored.

FTP provides an easy way to distribute files such as lecture notes and manuscripts; the files are simply placed on a computer for students and colleagues to download. For example, if an educator creates a program to demonstrate the Maxwell-Boltzmann equation in lecture, the program can be made available for students to download via FTP and run on their own computers. This procedure saves time by eliminating the need to make copies of floppy disks and distribute them in class.

Telnet

Telnet allows users to run programs on a remote computer. To use telnet, a user typically must have an account--namely, a userid and password--on the remote computer. Some computers, however, support guest telnet accounts that do not require a user to have these. Once connected, the local computer acts as the remote computer terminal, enabling the user to run applications, such as computational chemistry programs.

Telnet is useful if a chemist has more than one computer that he or she uses regularly. Suppose, for example, a chemist works part-time as a teacher at a college and part-time as a computational chemist in industry. Using telnet, the chemist can check the progress of computations on the industry computer while at college, and can respond to student e-mail inquiries on the college computer while at the industrial site.

Gopher

Five years ago, Internet users did not have access to graphical user interfaces and use of the world-wide web was not yet widespread. Gopher was the easiest way to get information via the Internet then, because it provided a text-only interface much like the table of contents of a book. By selecting the equivalent of a chapter, section, subsection and so on, a user could progressively narrow a search for information.

Because of its ability to transfer more than just text, the World Wide Web has mostly supplanted gopher. However, some information resources, such as the Journal of Chemical Education (**gopher://jchemed.chem.wisc.edu/**), are available in text-only format via gopher.

Electronic Discussion Groups

Discussions occur every day on the Internet in electronic discussion groups. There are thousands of discussion groups (**http://www.tile.net/tile/news/** and **http://www.tile.net./tile/listserv/**), some restricted to just a few people and others open to the entire world. Discussion themes, called threads, can be sedate, rational, and well-thought out, or they can be emotional, strident and fiery, just like face-to-face conversations.

In electronic discussion groups, e-mail is used in a one-to-many communication mode. A participant in a discussion group composes an electronic message and sends it to all members of the group at once. Similarly, the participant receives all mail sent to the group by other members. Electronic discussion groups lack the real-time information exchange that occurs in meetings or in conference calls, but they allow more rapid, convenient and wide-spread discussion than would occur with regular mail. There is also a degree of anonymity in electronic discussion groups. This anonymity enables some participants to say things they would not normally say in face-to-face meetings or on the telephone.

There are two kinds of electronic discussion groups; they differ in how they distribute e-mail messages. In one kind of group, called a newsgroup, e-mail is posted to a central computer, called a news server. To read or send mail to the group, a user connects to the news server with a program called a newsreader. In the other kind of group, usually called a listserver, e-mail is routed through a central computer and sent directly to the e-mail account of each group member. A newsgroup is analogous to a post office box; a user must walk down to the post office to check his or her mail. A listserver is analogous to having mail delivered directly to a home mailbox.

The two kinds of electronic discussion groups also differ in who usually has access to messages. In a newsgroup, anyone can post or read messages. In a listserver, only users who are allowed to join the discussion group can post or read messages, although many listserver discussion groups are open to anyone. Messages posted to a newsgroup are analogous to messages posted on a bulletin board in a public place, while messages posted to a listserver are analogous to messages posted on a bulletin board of a private club.

Audio and Video Conferencing

Audio and video can be transmitted and received in real time over the Internet. There are two common modes: a one-to-one mode, in which two users talk and see each other exclusively, and a one-to-many mode, in which a group of people exchange audio and video.

Because audio can be transmitted, the Internet can be used as a substitute for the telephone; however, the only advantage to this appears to be savings in long-distance phone charges, as there is a delay in the transmission of audio over the network. Communicating by video conferencing enables non-verbal forms of communication, such as facial expressions and body language, that are not communicated by voice alone. Although it will never replace face-to-face meetings, video teleconferencing can be used to build a school without walls so that students do not have to be in one place to attend and participate in lectures.

CONNECTING TO THE INTERNET

What is necessary to connect to the Internet? The requirements can be divided into hardware and software, which are discussed in the next sections. Hardware includes a computer as well as peripherals, such as a modem and its various interface cards. Software includes programs such as a World Wide Web Browser and Browser Helper Programs, described below.

Hardware

While virtually any computer can be used to connect to the Internet, a relatively powerful computer system should be used so that the computer does not become a bottleneck in playing multimedia files. This means a 486-class computer or better with at least 8 MB of memory, or a Macintosh with at least a 68030 processor and 6 MB of memory, or a PowerPC processor. The computer should have a graphical operating system (e.g., MacOS, Windows, Windows 95 or X Windows). A high-capacity disk drive, on the order of 1 GB or more, is useful if video and audio files are going to be downloaded. A sound card is necessary to play audio files and the audio part of video. If two-way teleconferencing is planned, a video camera, microphone, and video capture card are also necessary.

There are currently two common ways to physically connect a computer to the Internet: through ordinary phone lines, or through specialized wires or optical fibers (known by names such as T1, T3, and ATM). The specialized connections operate at much higher speeds, but are much more expensive. Connections such as these are usually found at institutions such as colleges, universities and larger companies. Access to these connections is normally controlled by the institutions' computer centers, which must be consulted to buy compatible hardware and software, and to connect a computer to the Internet.

From home, most users access the Internet via phone lines. Phone lines have the advantage of low cost and wide access, but are relatively slow. The connection between the computer and the phone line is through a modem, which converts the 0's and 1's (bits, or digits,) of computer language into the audio tones of a normal (analog) phone line. Current modems operate at a maximum transfer rate of 28.8 Kbps (1 bps = 1 bit/sec).

A digital phone line (integrated service data network, or ISDN) is an alternative to an analog phone line. ISDN lines are phone lines that transfer data digitally, as 0's and 1's, without translating them into audio tones. Because switching digital phone signals is faster than analog signals, data in digital format can be transferred much more quickly through the phone system than data transferred in audio format. The minimum data transfer rate of an ISDN line is 64 Kbps, and transfer rates of 128 Kbps are common. Digital ISDN lines are available in about 60-80% of the country, but are more expensive than analog lines. The local phone company should be contacted to install an ISDN line. In addition, an ISDN card may need to be installed in the computer.

It is important to consider the speed of an Internet connection when playing audio and video in real-time, and when downloading large files. The minimum speed for running real-time audio and video applications is 14.4 Kbps. Under optimal conditions, it takes 10-15 minute to access 1 MB of information with a modem operating at 14.4 Kbps.

Software
In addition to hardware, software applications need to be installed on the computer to access the Internet. One kind of software, the "driver" software, serves as an interface between the computer and Internet programs, and is required to make the computer Internet ready. Other kinds of software are required to browse the Internet. (For instructions on how to obtain the software listed in this section, see the section Freeware and Shareware Programs below. If Internet access is through a commercial on-line service, as described in the section Internet Service Providers and Online Services below, interface software will usually be provided free as part of the service.)

Driver Software
For Macintosh, the driver software is almost always MacTCP (**http://www.math.niu.edu/~behr/docs/mactcp.html**). System 7.5 and later versions of the Macintosh operating system come with MacTCP. For earlier versions of the Macintosh Operating System (MacOS), the software is available from Apple or from a software distributor (e.g., **http://www.warehouse.com/**). Some books, such as the Internet Starter Kit for Macintosh by Adam Engst, include MacTCP for less than the price of MacTCP alone.

Additional software is necessary if the Internet connection is by modem or ISDN. There are two standards in transmitting data through phone lines: Serial Line Internet Protocol (SLIP) and Point-to-Point Protocol (PPP). PPP is similar to SLIP, but offers additional checking for data integrity. For Macintosh, a software application for SLIP connections is InterSlip (available at the anonymous FTP site **ftp://ftp.intercon.com/InterCon/sales/Mac/Demo_Software**). A software application for PPP connections is MacPPP (**ftp://wuarchive.wustl.edu/systems/mac/umich.edu/util/comm/macppp2.12sd.sit**.

hqx). Commercial versions of these products, which have the advantage of technical support, are also available.

A computer running Windows 95 or OS/2 is Internet ready and does not need any additional software. For Windows and Windows for Workgroups 3.x, Windows Sockets driver software (winsock) is necessary. While in principle winsock programs from different manufacturers should be interchangeable, in practice they are not, and some applications need a winsock program from a particular manufacturer. A popular winsock for Windows is Trumpet Winsock (**http://www.trumpet.com.au/**). Microsoft offers a free winsock for users of Windows for Workgroups (**ftp://ftp.microsoft.com/peropsys/windows/public/tcpip/wfwt32.exe**).

Many new Internet applications for Windows, such as the World Wide Web Browsers described in the next section, are written in 32-bit code for increased performance. To run these applications on Windows and Windows for Workgroups 3.x, which are 16-bit operating systems, Win32s, available from Microsoft (**ftp://ftp.microsoft.com/softlib/mslfiles/pw1118.exe**), may need to be installed. Windows 95 is a 32-bit operating system, and does not need Win32s.

Installation of the driver software is usually a matter of dragging files to the system folder and various subfolders (Macintosh) or running a setup program (Windows). Configuration of the driver software varies from product to product, but usually involves specifying Internet Protocol (IP) addresses (e.g., 128.226.1.2) for the computer, for a gateway computer, and for a domain name server (DNS) computer, as well as a name for the computer and its domain. To maintain order on the Internet, these names and numbers should be assigned by the network system administrator of the institution or by the Internet service provider.

World Wide Web Browsers
Once the computer is Internet ready, the single most important program to install is a World Wide Web browser. There are many browsers to choose from and the number of features supported by them change rapidly. As of May, 1996, the best browser in terms of number of features is Netscape Navigator (**http://home.netscape.com/**), followed by Mosaic from the National Center for Supercomputing Applications (NCSA) (**http://www.ncsa.uiuc.edu/**). Microsoft has recently developed The Internet Explorer (**http://www.microsoft.com/windows/ie/explorer.htm**), but this browser runs only on the Windows 95 operating systems. For users at educational institutions, the latest versions of these browsers can be obtained free of charge at **http://home.netscape.com/comprod/mirror/index.html**, **http://www.ncsa.uiuc.edu/SDG/Software/Mosaic/NCSAMosaicHome.html**, and **http://www.microsoft.com/windows/ie/explorer.htm**. A commercial version of Netscape Navigator (**http://home.netscape.com/netstore/index.html**) is also available and has the advantage of technical support.

The most recent World Wide Web browsers not only access information in WWW format (known as HTML, or hypertext markup language), but also have such built-in features as e-mail, FTP, telnet, gopher, newsreader programs, and video and image display capabilities. Additional features continue to be built into browsers, such as the ability to run custom programs within the browser. These features make the World Wide Web browser a "command central" for access to the Internet. Because they contain so many additional features, World Wide Web browsers are *the* way to access information on the Internet. Additional software programs will soon be unnecessary.

World Wide Web browsers need to be configured for users to access newsgroups and to access electronic mail and participate in electronic discussion groups. The configuration involves specifying the IP address or name of the news server or mail server. The computer centers of academic and commercial institutions can provide these IP addresses or names. In the Netscape Navigator and NCSA Mosaic browsers, these IP addresses or names are specified under Preferences in the Option menu. World Wide Web browsers from some on-line services, described in the section Internet Service Providers and Online Services, are proprietary and have different configuration methods. These companies can provide the addresses of their news servers and mail servers.

The parallel development of the WWW and of Web browsers has transformed the Internet into a content-rich information source. It no longer matters what kind of hardware or software is used. These factors are now irrelevant. Users can access the information on the Internet with virtually any machine, running any operating system.

Browser Helper Programs
Although the number of features supported by World Wide Web browsers is increasing rapidly, there are still some additional "helper" applications that may need to be installed to take full advantage of the Internet. Two of these are a telnet application (for WWW links that begin with telnet://) and an application for displaying video. (The files described in this section can be downloaded using the procedures described in the section, Freeware and Shareware Programs.)

For Windows, a free telnet application for Windows is ewan (**ftp://ftp.cica.indiana.edu/pub/pc/win3/winsock/**). A Macintosh version of the telnet application is NCSA Telnet (**http://www.ncsa.uiuc.edu/SDG/Software/Brochure/MacSoftDesc.html#MacTel net**). Once a telnet program is downloaded and installed, the name of the directory, or folder, of the telnet program should be defined in the World Wide Web browser. For Netscape Navigator and NCSA Mosaic, this is specified under Preferences or Mail and News Preferences in the Options menu. For further details, consult the installation manual or browser help files.

The two most common video formats found on the Internet are QuickTime and MPEG. To download and view video, software needs to be installed that will play these formats on a particular computer. For Macintosh, QuickTime is included with Versions 6 and 7 of the Mac operating system. For Windows, an early version of QuickTime for Windows is available for free (**http://www.ncsa.uiuc.edu/SDG/Software/WinMosaic/Viewers/qt.htm**), and newer versions are available for a nominal fee from Apple (**http://quicktime.apple.com/**) or software distributors. An application to play MPEG video on the Macintosh is Sparkle (**ftp://ftp.ncsa.uiuc.edu/SDG/Software/WinMosaic/Viewers/mpg.htm**). For Windows, two MPEG video applications are mpegplay (**http://www.ncsa.uiuc.edu/SDG/Software/WinMosaic/Viewers/mpeg.htm**) and vmpegwin (**ftp://ftp.cica.indiana.edu/pub/pc/win3/desktop/vmpeg12a.zip**).

An increasingly common way to play MPEG video is through hardware rather than through software. The hardware usually takes the form of a special board or chip placed in the computer. The advantages of hardware playback are faster, smoother video and the ability to display video on the full screen of a computer monitor.

Once a video playback program has been downloaded and installed, the name and location, i.e., the directory or folder, of the program needs to be entered into the World Wide Web browser. For Netscape Navigator and NCSA Mosaic, this location is specified under Preferences or General Preferences of the Options menu. Browsers use a convention called Multipurpose Internet Mail Extensions (MIME) to transfer data other than text through the Internet. Browsers need to be configured so that they channel non-text data to the appropriate translator. The Netscape Navigator and NCSA Mosaic come preconfigured to recognize QuickTime and MPEG formatted video. Specifying the location of the QuickTime and MPEG playback programs as described above is then sufficient to play Internet video. Configuring a browser to start a particular program involves specifying the MIME type, subtype and file extensions, and the name and location of the "helper" application. For QuickTime, the type/subtype is video/QuickTime, and typical file extensions are qt and mov. For MPEG, the MIME type/subtype is video/mpeg, and typical file extensions are mpeg, mpg and mpe. MIME types and "helper" program locations are specified in Netscape Navigator and NCSA Mosaic under Preferences in the Option menu. The installation manual or help files of a particular browser should be consulted for further details.

Freeware and Shareware Programs
A browser such as Netscape Navigator or NCSA Mosaic, together with separate telnet and video applications, is sufficient to find and retrieve most of the information from the Internet. However, there are several additional kinds of applications that are useful. For example, although browsers can download files using their internal FTP, they cannot upload files. To be able to upload files, a separate FTP application is necessary. Suggested FTP applications and sources are listed in the table below.

To download a file from a World Wide Web browser, the URL of the FTP site is entered and the file is downloaded to disk. (If you are using the on-line version of this guide, you can click directly on the URL to download files.) For example, to download the FTP application for Macintosh, Fetch, **ftp://ftp.dartmouth.edu/pub/mac/Fetch_2.1.2.sit.hqx** would be entered as the URL.

Files that are transferred via FTP are usually in compressed format. For PCs, the most common compression format is the zip format; these files are denoted by the .zip file name extension (a file name extension is the letters that occur after the period "." in a file name). PC files are uncompressed by running the shareware program pkunzip (**http://www.pkware.com/**). For Macintosh, two common compression formats are Stuffit (.sit) and CompactPro (.cpt). Files are sometimes also encoded into BinHex format (.hpx) or MacBinary format (.bin). All of these formats can be uncompressed and decoded using the program Unstuffit-Expander (**ftp://ftp.ncsa.uiuc.edu/Mosaic/Mac/Helpers/**). Some files contain a mini uncompression routine that runs automatically when the compressed file is opened; these are denoted by the file name extension .exe on the PC and .sea on the Macintosh.

Applications for Macintosh Platform

Type of Program	Application	Source
audio	soundapp	ftp://wuarchive.wustl.edu/systems/mac/umich.edu/sound/soundutil/soundapp1.51.cpt.hqx
compression	DropStufft	ftp://wuarchive.wustl.edu/systems/mac/umich.edu/util/compression/dropstuff3.52.sea.hqx
document viewer	Acrobat	ftp://ftp.adobe.com/pub/adobe/Applications/Acrobat/Macintosh/ACROREAD.MAC.hqx
FTP	fetch	ftp://ftp.dartmouth.edu/Fetch_2.1.2.sit.hqx
image	GraphicConverter	ftp://ftp.ncsa.uiuc.edu/Mosaic/Mac/Helpers/graphic-converter-22.hqx
e-mail	eudora	http://www.qualcomm.com/ProdTech/quest/EudoraLight.html
molecule viewer	Rasmol	ftp://colonsay.dcs.ed.ac.uk/export/rasmol/rasmac.hqx
teleconference	CUSeeMe	http://cu-seeme.cornell.edu/ and http://www.wpine.com/cuseeme.html

Applications for PC Platform

Type of Program	Application	Source
audio	wplany	http://www.ncsa.uiuc.edu/SDG/Software/WinMosaic/Viewers/wplany.htm
compression	WinZip	http://www.winzip.com/
document viewer	acroread	ftp://ftp.adobe.com/pub/adobe/Applications/Acrobat/Windows/ACROREAD.EXE
	wordview	http://www.microsoft.com/msoffice/freestuf/msword/download/viewers
	gsview	http://www.cs.wisc.edu/~ghost/gsview/
FTP	ws_ftp	ftp://ftp/cica.indiana.edu/pub/

image	LViewPro	pc/win3/winsock/ws_ft.zip ftp://uiarchive.cso.uiuc.edu/pub/ systems/pc/simtel/win3/graphics/ lviewp1b.zip
e-mail	pceudora	http://www.qualcomm.com/ ProdTech/quest/EudoraLight.html
molecule viewer	Rasmol	ftp://colnsay.dcs.ed.ac.uk/ export/rasmol/raswin.zip
teleconference	CuSeeMe	http://cu-seeme.cornell.edu/ and http://www.wpine.com/cuseeme.html

There are thousands of additional freeware and shareware programs available at anonymous FTP sites. The easiest way to search for software is to use the Virtual Software Library "front desk" at Oakland University (**http://www.acs.oakland.edu/cgi-bin/vsl-front**) which allows users to search 22 of the largest FTP sites by title and keyword. NASA offers another way to search of FTP sites (**http://www.lerc.nasa.gov:80/archiplex/doc/form.html**).

Commercial Software
The examples of software listed in the previous section are examples of freeware or shareware. Freeware can be used without charge, while shareware has a limited period of free usage, after which the software must be purchased. Note that freeware and shareware are not necessarily public domain software; the author of the software usually retains copyright and there are usually restrictions about copying and distributing the software.

Commercial versions of freeware and shareware programs are available. The advantages of commercial software are usually better technical support, fewer bugs, more robust operation, and a more consistent stream of updates. The disadvantage, of course, is that commercial software costs more.

Internet Service Providers and On-line Services
Many academic and commercial institutions support access to the Internet through SLIP and PPP phone lines, so that a chemist employed at one of these institutions can connect to the Internet from home by calling one of these lines. In the absence of such support, a chemist at home can still connect to the Internet through an Internet service provider or on-line service.

On-line services offer Internet access as part of a larger package of services. Popular on-line services include America OnLine (**http://www.aol.com**), CompuServe (**http://www.compuserve.com**), Prodigy (**http://www.prodigy.com**) and, most recently, the Microsoft Network (**http://www.msn.com/**), as well as General Electric's GEnie (**http://www.genie.com/**) and Apple's eWorld (**http://eweb02.online.pple.com/webcity/eworld/**). The pros and cons of the various on-line services are reviewed often in various magazines (e.g., **http://www.zdnet.com/~pccomp/features/internet/online.html**) but the features and service fees of each company change so rapidly that printed reviews quickly become out of date. On-line services usually have a local phone number to call, offer modem access at 14.4 Kbps and 28.8 Kbps, and distribute their own software to connect to and use their services. Some of these companies are upgrading to digital ISDN phone lines in selected cities to offer 64 Kbps or 128 Kbps service. In addition to Internet access, the large on-line services offer proprietary services such as the ability to make airline reservations on-line, and access to exclusive on-line interviews with celebrities.

Internet access can also be obtained through Internet service providers that provide Internet access without the additional features of on-line services. Internet tools that these companies typically provide include e-mail, FTP, telnet, and WWW access. Internet service providers can be small local companies, or larger regional or national companies. For Internet access only, it is probably cheaper and faster to use an Internet service provider rather than an on-line service. Internet service providers are listed on-line at (**http://www.gnn.com/gnn/wic/netinfo.63.html**) and in the phone book.

SEARCHING THE INTERNET
The amount of information available on the Internet is staggering. It was estimated to grow to 50-100 terabytes (1×10^{12} bytes) by the end of 1995. Finding a particular needle of information in the haystack that is the World Wide Web may at first seem a daunting task, but there are now many search tools that enable the chemist to locate information with ease.

Finding People
If a chemist needs to talk to someone on the phone and does not know the person's phone number, he or she can look in a phone book or call directory assistance. The Internet directories are not yet as organized as the telephone directories of a phone company; therefore, it may take some time to find an person's e-mail address. Sometimes, the quickest way to find someone's e-mail address is to call the person on the phone.

The most current information about how to find someone's e-mail address can be found at **http://www.cis.ohio-state.edu/hypertext/faq/usenet/findingaddresses/faq.html**. Many universities offer an Internet directory of faculty and staff; a collection of these directories is maintained at Notre Dame (**gopher://gopher.nd.edu:70**). A directory of chemists has been started (**http://hackberry.chem.niu.edu:70/0/ChemDir/index.html**) and currently contains about 500 listings. Search procedures using the UNIX-based utilities **finger** and **whois** (**http://www.nova.edu/Inter-Links/phone.html**) can be used to locate someone if the person's institution is known. Many Internet users maintain a "home page" on the WWW; indexes to these can be found at the University of Texas (**http://www.utexas.edu/world/personal/index.html**).

In addition to institutional e-mail addresses, some chemists have e-mail addresses through on-line services or Internet service providers. Places to start searching for these addresses are Four11 (**http://www.four11.com**) and LookUP! (**http://www.lookup.com/search.html**). Together, these two sites offer e-mail addresses of more than 1.5 million Internet users. Both are commercial services, but currently offer free searching of their databases. For reasons of privacy, on-line services such as CompuServe and Prodigy do not make available directories of their members to people who do not subscribe to their services. The more general search procedures described in the next section can also be used to search for e-mail addresses.

Finding Information
The millions of pages of information on the World Wide Web are continuously indexed by specialized programs (known as webcrawlers, spiders or robots) that automatically visit sites on the Internet and collect information about what is available there. These enormous databases can be searched by topic or keyword.

A collection of different ways to search the Internet can be found at Nexor (**http://pubweb.nexor.co.uk/public/cusi/cusi.html**). The most comprehensive database is Lycos (**http://lycos.cs.cmu.edu/**), which, in January, 1996, contained references to 10 million WWW pages, representing 91% of the total Web pages in existence at that time. Another popular database is InfoSeek (**http://www2.infoseek.com/**), available as a button in the Netscape Navigator browser. An index to over 8 billion words in WWW pages can be found at Alta Vista (**http://altavista.digital.com/**).

Many sites contain databases of information about particular topics. These databases can be searched using a Wide Area Information Server (WAIS) over the WWW (**http://www.wais.com/**). A database of postings to newsgroups can also be searched (**http://www.dejanews.com/forms/dnq.html**).

There are collections of WWW pages that are indexed according to various categories. The original collection is Yahoo (**http://www.yahoo.com/**). The Whole Internet Catalog (**http://nearnet.gnn.com/wic/**) is also a popular place to browse by topic.

There are many WWW sites of interest to chemists. For example, a gopher site containing MSDS is available at the University of Utah (**http://gopher://atlas.cem.utah.edu/11/MSDS**). Two on-line versions of the periodic table are WebElements (**http://www.cchem.berkeley.edu/Table/index.html**), which sports an isotope and element percentage calculator, and the MPEG Periodic Table (**http://huckel.cm.utexas.edu/mpegtable.html**), which features MPEG video of reactions of the elements.

Comprehensive lists of sites of interest to chemists, also called "meta-sites," contain primarily lists of links to other sites. Some meta-sites for the chemistry educator are:
        http://www.rpi.edu/dept/chem/cheminfo/chemres.html
        http://www-hpcc.astro.washington.edu/scied/chemistry.html
        http://www.anachem.umu.se/eks/pointers.htm

Some sites that contain more general information about chemistry are:

http://www.indiana.edu/~cheminfo
http://www.chem.ucla.edu/chempointers.html
http://www.yahoo.com/Science/Chemistry
http://www.shef.ac.uk/~chem/chemdex
http://chemistry.rsc.org/rsc

There are also a large number of sites containing general information about the Internet and the WWW; some of these are:
http://ds.internic.net/ds/dsdirofdirs.html
http://www.globalcenter.net/gcweb/tour.html
http://www.hw.ac.uk/libWWW/irn/irn13/irn13.html
http://sunsite.unc.edu/~boutell/faq/
http://www.lib.umich.edu/chhome.html
http://www.internic.net/ds/dsdirofdirs.html

THE INTERNET IN CHEMISTRY EDUCATION

The Internet offers instructors ways to try new and exciting ways of teaching chemistry. Some of these ideas are listed in the sections below, and more experimental approaches are listed in the section The Future of the Internet, below.

Course Material

One use of the Internet is to distribute course material such as syllabi and lab instructions. The University of Texas has started a collection of such material (**http://wwwhost.cc.utexas.edu/world/instruction/ch/**). Instructors can quickly and easily update or modify the information, and students can access the updated course materials from their dorm rooms, off-campus apartments, or from anywhere there is Internet access.

There are a number of high quality chemistry tutorials on the WWW (**http://www.ch.ic.ac.uk/GIC/**) that can be used to supplement the lecture and textbook. Two particularly good sites are Virginia Polytechnic Institute and State University (**http://www.chem.vt.edu/chem-ed/vt-chem-ed.html**) and the University of California, San Diego (**http://www-wilson.ucsd.edu/**). These sites contain databases of multimedia for the core undergraduate courses. In addition to text readings and problems, instructors could assign perusal of particular sections of these and other Internet sites as homework.

Undergraduate and graduate seminar courses should include instruction in how to use the Internet because the Internet represents a new way to conduct research and access information. Many university libraries have developed WWW interfaces to their catalogs and databases, making it convenient to search for chemical information from the laboratory, office, or home. While most commercial information services charge for access to their databases, CARL Corporation offers free access to a large number of databases (**telnet://database.carl.org/**); these databases can be used in real time to demonstrate to a seminar class how to search the scientific literature on-line.

The Internet can be used to supplement lectures. Lecture rooms are often equipped with computers and LCD display panels or video projection equipment, so that the instructor can project lecture notes and supplements onto a large overhead screen. With some additional investment, the lecture room can be wired for the Internet to further supplement lectures. For example, a protein secondary structure prediction by neural networks (**http://www.cmpharm.ucsf.edu/~nomi/nnpredict.html**), could be done in real-time to give students the flavor of what a practicing biochemist might do. The Brookhaven Protein database (**http://www.pdb.bnl.gov/cgi-bin/browse**), can be used to display and rotate structures in real-time using a molecular viewer such as Rasmol (**ftp://colonsay.dcs.ed.ac.uk/export/rasmol/**). As always, however, there should be contingency plans in case the computer system fails or the Internet connection becomes too slow. Many lecture supplements obtained from the Internet, such as video of the SN^2 reaction (**http://ils.unc.edu/dopamine/images/sn2.mpg**) or the latest information on ozone depletion (**http://www.epa.gov/docs/ozone/**), are best retrieved before the lecture to avoid interruptions in the lecture while files are downloading.

The Internet has also been used to teach courses electronically (**http://www.gnacademy.org/**). The most extensive courses offered on the Internet are a course on the C++ programming language (**http://info.desy.de/gna/html/cc/index.html**) and a course in materials science (**http://vims.ncsu.edu**). One of the instructors of the C++ course has offered a retrospective on the course (**http://uu-gna.mit.edu:8001/uu-gna/text/cc/papers/2nd_conf.html**). The C++ course made use of a MOO (Multi-user dungeon Object Oriented). Originally developed for the game Dungeons and Dragons, a MOO creates a world in which objects can be manipulated by text commands. Students and instructors can interact using text-based terminals. Electronic chemistry courses such as Computational Chemistry for Chemistry Educators (**http://www.mcnc.org/HTML/ITD/NCSC/ccsyllabus.html**) have started to appear. It is not clear at this stage whether these kinds of courses will replace traditional methods of delivering chemistry instruction or serve as a supplement to existing courses.

Video teleconferencing through the Internet represents a new way to deliver distance learning. A low-tech way to do this without slowing down the Internet too much is to use the program CuSeeMe (**http://cu-seeme.cornell.edu/**) which allows video transfer through the Internet, even through a 14.4 Kbps modem. Installation of CuSeeMe on Internet-ready Windows and Macintosh computers allows users to tune in to broadcasts on the Internet, such as real-time video from the space shuttle. By attaching a video camera and microphone to his or her computer, an instructor could broadcast lectures and office hours over the Internet for remotely-located students to receive to students.

With the current state of this technology, the quality of real-time video delivered over the Internet is inferior to that delivered by video broadcasts such as closed-circuit or cable television. One way to improve the quality of real-time instruction is to add a program such as Timbuktu (**http://www.farallon.com/www/gen/software.html**) or Collage (http://www.ncsa.uiuc.edu/SDG/Software/Brochure/Overview/MacCollage.overview.html). These programs allow an instructor to change a computer screen containing, for example, a spreadsheet or image and have the change broadcast immediately to students. In this way students would get a real-time development of material presented on the computer screen to go along with the real-time audio and video.

Electronic Discussion Groups
Electronic discussion groups can be used within the context of a course. Group membership can be restricted to just students taking a particular course. Students can post questions or comments to the group and anyone in the group can respond, whether it be an instructor, teaching assistant, or fellow student; thus, the the learning environment becomes more collaborative. For large enrollment classes, about 5-10% of the students participate actively in the discussions, but a much larger fraction of students participate passively. These electronic study groups should be set up in consultation with the computer network administrator.

Office hours can be held online via electronic discussion groups. Instructors can be available on-line during a particular time. Students can, then, post questions to the discussion group and have them answered. A log of the discussions could be kept for access by other students, cutting down on the number of times an instructor has to answer the same questions. Although this method frees students from having to congregate in one room for review sessions, the rate of information exchange is much slower than in conversation.

Instructors can join other electronic discussion groups to keep abreast of the latest developments and trends in chemistry education. For example, an educator may have a problems with a particular laboratory experiment. By posting a query to an electronic discussion group, the instructor may receive a reply from another instructor who has had the same problem and can offer a solution. Please see the list of various discussion groups below.

There are two popular kinds of electronic discussion groups: newsgroups and listservers. In a newsgroup, electronic discussion messages are stored on a central computer, called a news server. Anyone with Internet access and newsreader software can view the messages posted to a newsgroup. In a listserver, messages are automatically routed to the electronic mailboxes of all members of the group. Users must subscribe to a listserver to be able to view the messages.

Listserver discussion groups are subscribed to by sending an e-mail message to the listserver address, shown in the table below. (This address is different from the address used to send mail to the discussion group.) The e-mail message should contain a single line
     subscribe   name-of-list   your-name
where name-of-list is the name of the discussion group to subscribe to, and your-name is your name. For example, to subscribe to the Chemistry Education list (chemed-l), an e-mail message would be sent to listserv@uwf.cc.uwf.edu; the body of the message would be

> subscribe chemed-l James A. Dix

To unsubscribe to a listserver group, an e-mail message is sent to the listserver address containing the single line message

> unsubscribe name-of-list

For example, to unsubscribe to the Chemistry Education list, an e-mail message would be sent to listserv@uwf.cc.uwf.edu; the body of the message would be

> unsubscribe chemed-l

Once the subscribe message is received and approved by the listserver, a message is usually sent to the subscriber describing the policies and procedures of the discussion group. To actually participate in the electronic discussions, messages should be sent to the address of the discussion group, not of the listserver. For example, to participate in the Chemistry Education discussion (chemed-l), messages should be sent to chemed-l@uwf.cc.uwf.edu, not listserv@uwf.cc.uwf.edu. All messages sent directly to the group are received by the centralized computer, or server, and then routed to individual members' e-mail accounts. Popular groups generate many e-mail messages every day; disk space may fill up quickly if mail is not discarded regularly.

A list of listserver groups of interest to chemical educators, along with the addresses to send subscribe or unsubscribe messages, is given below. (If you are using the on-line version of this guide and are using a browser supporting e-mail, you can click directly on the addresses to send subscribe messages.)

List name	Listserver address	Discussion group	Topic
bchfac-l	listserv@unlvm.unl.edu	bchfac-l@unlvm.unl.edu	Biochem. Faculty (selective)
chemchat	listserv@uafsysb.uark.edu	chemchat@uafsysb.uark.edu	Student & Professional Discussions
chemclub	listserv@umslvma.umsl.edu	chemclub@umslvma.umsl.edu	Chemistry Club (selective)
chemcom	listserv@ubvm.cc.buffalo.edu	chemcom@ubvm.cc.buffalo.edu	Chemistry in the Community
chemconf	listserv@umdd.umd.edu	chemconf@umdd.umd.edu	Confs. on Chem. Research and Educ.
chemcord	listserv@umdd.umd.edu	chemcord@umdd.umd.edu	General Chemistry Coordinators
chemdisc	listserv@umdd.umd.edu	chemdisc@umdd.umd.edu	Chemistry Conference Discussion
chemed-l	listserv@uwf.cc.uwf.edu	chemed-l@uwf.cc.uwf.edu	Chemistry Education Discussion
chemistrytm	chemistrytm-request@dhvx20.csudh.edu	chemistrytm@dhvx20.csudh.edu	Chemistry Telementoring
chemlab-l	listserv@beaver.bermidji.msus.edu	chemlab-l@beaver.bermidji.msus.edu	Chemistry Laboratory Discussion
chemweb	listserver@ic.ac.uk	chemweb@ic.ac.uk	WWW in Chemistry
chminf-l	listserv@iubvm.ucs.indiana.edu	chminf-l@iubvm.indiana.edu	Chemical Info. Sources Discussion
cicourse	listserv@iubvm.indiana.edu	cicourse@iubvm.indiana.edu	Chemical Information Courses
cqshare	listserv@uwf.cc.uwf.edu	cqshare@uwf.cc.uwf.edu	Chemistry Question Sharing
microscale-1	microscale-1-request@merrimack.edu	microscale-l-request@merrimack.edu	Microscale Chemistry
safety	listserv@uvmvm.uvm.edu	safety@uvmvm.uvm.edu	Laboratory Safety

For newsgroups, there is no formal subscription procedure. Instead, a newsreader program is used to read the news. The Netscape Navigator and NCSA Mosaic WWW browsers have newsreaders built in. To read the news of sci.chem, for example, the address news:sci.chem would be entered in the browser as the URL. This assumes that the browser has been configured with the correct address of the news and e-mail servers. The following newsgroups are of interest to chemistry educators:

newsgroup	topic
sci.chem	Chemistry and related sciences
sci.chem.labware	Chemical laboratory equipment
bionet.molec-model	The physical & chemical aspects of molecular modeling

Network etiquette ("netiquette") should be followed when participating in electronic discussion groups. Many groups have lists of Frequently Asked Questions (FAQs) which provide answers to commonly-asked questions. When joining a group for the first time, it is a good idea to check the FAQs. A list of FAQs for newsgroups can be found at Ohio State (**http://www.cis.ohio-state.edu/hypertext/faq/usenet/top.html**). Listserver groups usually have an administrator who can give additional information about the discussion. The introductory messages sent to subscribers give the policies and procedures of the group. For example, the group may have an unsubscribe procedure that is different from the normal "unsubscribe listname" message. Also, it is best to "lurk" on the list for a while (i.e., read only) to learn the scope, tenor and tone of the discussion before posting messages.

On-line Journals
An excellent example of what can be done with on-line publishing is the WWW site for the Journal of Biological Chemistry (**http://www-jbc.stanford.edu/jbc/**). The full text of articles is available on-line for free, as are figures, tables and references with links to the Medline database. All text can be searched on-line.

The American Chemical Society is translating its many journals into on-line format. But access to ACS journals is usually not free. For example, to access supporting material for articles published in the Journal of the American Chemical Society, a JACS subscription number must be entered into the Web browser (**http://pubs.acs.org/supmat/jacsat/forms/v117form/v117-ja12001.html**). On-line access to ACS material, originally through STN and modem, has been updated to include Internet access (**http://www.cas.org/ONLINE/online.html**). Charges for STN include connect time, number of articles accessed, and number of articles downloaded, with academic discounts for off-peak times. The ACS has produced SciFinder (**http://info.cas.org/ONLINE/online.html**) to provide easy access to on-line journals; the software is marketed to large research firms. A personal version of SciFinder is being considered, as is WWW access.

The ACS Division of Chemical Education newsletter is on-line (**http://jchemed.chem.wisc.edu/cheds.html**). The Journal of Chemical Education has a WWW site (**http://www.utexas.edu/ftp/cons/jchemed/**) containing limited information about the journal and JCE:Software programs. There is also a gopher server containing some information about the Journal of Chemical Education (gopher://jchemed.chem.wisc.edu/). No full on-line equivalent of the journal is available. The Journal of Chemical Education, in an editorial in the November, 1995 issue, stated that it would be "intellectually and economically unacceptable to produce only a direct copy" of the journal.

A new on-line, peer-reviewed chemistry education journal has been announced: The Chemical Educator (**http://CHEDR.IDBSU.EDU**). This journal will be totally electronic and have no counterpart in the print medium. Access to The Chemical Educator, published by Springer-Verlag starting March 1, 1996, will be free for an introductory period, followed by a "very nominal yearly subscription fee."

Electronic Conferences

Electronic conferences are common on the WWW. In this type of conference, presenters create "posters" containing text, graphics, video, and other content in WWW format, then post the content either on their own computer or on a central conference computer. During the actual dates of the conference, "attendees" view the electronic posters content on the WWW, and submit and respond to comments about the electronic information. Conference proceedings consist of the WWW conference sites along with attendees' comments. One of the first chemistry electronic conferences was about trends in organic chemistry (**http://www.ch.ic.ac.uk/ectoc**). As part of ChemConf 96, an on-line symposium, New Initiatives in Chemical Education, is scheduled for June 3 to July 19, 1996 (**http://www.wam.umd.edu/~toh/ChemConf96.html**).

Creating and Publishing WWW Content

The programming language of the WWW is the HyperText Markup Language, or HTML (**http://www.w3.org/hypertext/WWW/Tools/Filters.html**). The Beginners Guide to HTML (**http://www.ncsa.uiuc.edu/General/Internet/WWW/HTMLPrimer.html**) describes the HTML language in detail. HTML is a markup language, which means that text is given tags that instruct the Web browser to, for example, display text in italics or link to another Web resource when a text link is selected. (If you are using the on-line version of this guide, you can view the HTML for this page by viewing the source code of the document. To do this using Netscape Navigator, select Source from the View menu, or if you are using NCSA Mosaic, select View Source from the File menu.)

Fortunately, chemists do not need to know HTML to create WWW content. For example, the on-line version of this guide was created with Microsoft Word 6.0 using HTML Author templates (**http://www.salford.ac.uk/docs/depts/iti/staff/gsc/htmlauth/summary**). An extensive list of templates and conversion programs is given at CERN (**http://www.w3.org/hypertext/WWW/Tools**). Several on-line manuals give advice on the design of WWW pages. For example, the Yale Web Style Manual (**http://info.med.yale.edu/caim/StyleManual_Top.HTML**) offers tips on what it considers to be good WWW page design. A meta-site providing advice on HTML creation is available at Indiana University (**http://www-slis.lib.indiana.edu/Internet/programmer-page.html**). With these tools, almost anything stored on a computer, ranging from course outlines to educational software, can be published on the WWW. A collection of useful material on the WWW has been started (**http://www.ch.ic.ac.uk/GIC/contribute_form.html**).

To publish on the WWW, files must be placed on an computer running Hypertext Transport Protocol (HTTP) server software. The institutional computer center might have computers already configured that can be used to publish material. Instructors could then give the URL of the site to students as part of the ancillaries to a class. A HTTP server can also be set up on a personal Macintosh (**http://www.biap.com/**) or Windows (**http://www.city.net/win-httpd/**) computer, allowing the instructor greater freedom and accessibility in controlling the content of the site. Setting up a WWW server requires some technical expertise and coordination with the institutional network administrator.

General Interest
Chemists do not live by chemistry alone, and the WWW provides ways to take a refreshing break from studying, administration, and lecture preparation, and so on. A highly selective list of general interest sites is

Site Content	**Site WWW address**
news	http://www.enews.com/
weather	http://wx3.atmos.uiuc.edu/
sports	http://www.sportsnetwork.com/
home shopping moves onto the Internet	http://www6.internet.net/
stock market prices	http://www.secapl.com/cgi-bin/qs
CIA fact book	http://www.odci.gov/cia/publications/95fact/index.html
The Scientist magazine	http://www.internic.net/the-scientist/WWW/welcome.html
Science magazine on-line	http://www.aaas.org/science/science.html
cool site of the day	http://cool.infi.net/
Time-Warner site	http://www.pathfinder.com/
on-line Journal of Irreproducible Results	http://www.ugcs.caltech.edu/~maron/miniair/archive.html
modern day alchemists	http://www.colloquium.co.uk/alchemy/home.html
virtual tour of the world	http://wings.buffalo.edu/world/vt2/

ADVANTAGES AND DISADVANTAGES OF THE INTERNET
It is easy to become swept up in the hype surrounding the Internet. The Internet will not solve all communication problems nor will it alleviate world hunger. Chemists should be aware of the unique issues in using the Internet to teach and learn chemistry. Some of these issues are discussed in the following sections.

Security
Data sent over the Internet is not secure without encryption. Like a wiretap on a telephone, a "snooper" affixed to an Internet cable can monitor and decode Internet traffic. Sensitive information such as credit card numbers, student IDs and grades should not be transmitted through the Internet without secure encryption. Many

institutions have separate, secure wires to transmit such information. A common encryption is Pretty Good Privacy (PGP; **http://dcs.ex.ac.uk/~aba/pgp/**). The latest versions of NCSA Mosaic and Netscape Navigator support encryption without the use of external programs.

Reliability
The ease of publishing information on the Internet and the lack of critical review mean that information retrieved from the Internet must be evaluated carefully. Periodically, pranksters post to Internet discussion groups the dangers of the new chemical, dihydrogen oxide, and start a flurry of paranoid postings from chemically-challenged Internet users.

Electronic information is ephemeral. A lightning storm, lack of funding, and pernicious meddling can all take a computer off-line and deny access to information. Some Internet sites take these factors into account by providing backup power supplies, multiple copies of disks, and so on. Nonetheless, contingency plans are a necessity for any important lecture or presentation using a live Internet connection.

Information Overload
The amount of information on the Internet, coupled with the ease of accessing it and jumping from site to site on the WWW, makes it easy to become lost. It helps to focus when navigating the Internet and maintain a path toward a particular goal. A particularly useful navigational feature of World Wide Web browsers is the bookmark or history function. These features provide landmarks on a particular journey through the WWW and allow users to retrace their steps easily.

A larger question is whether the large amount of information on the Internet is good for chemistry education. Students may become bogged down with the details of chemistry and lose sight of larger, unifying concepts. Instructors should keep this in mind, and try to tie the loose bits of electronic information together in lectures and office hours.

Copyright Issues
Copyright has a long history in the United States, and guidelines for fair use of copyrighted material have been developed for traditional media such as print, video and audio. Because it is so easily copied and distributed, media on the Internet pose a unique set of issues (**http://www.benedict.com/**). There are no agreed-upon guidelines for the fair use of copyrighted materials. President Clinton has proposed new legislation to cover copyright of electronic media (**http://www.mecklerweb.com/simba/95090501.htm**).

A reasonable approach to take is to assume that everything on the Internet is copyrighted, whether explicitly or implicitly. Permission from the author should be obtained if, for example, video is downloaded and distributed to a class. On the other hand, the URL of the site can be freely given to the class, and each student could view the video without violating copyright law. It would also be reasonable to download a file and store the file on a computer for display during lecture, as long as the file was not distributed or sold for profit.

THE FUTURE OF THE INTERNET

Given growth estimates of Internet users, it appears that in less than a decade, there will be more Internet users than there are human beings on earth! The actual number of Internet users is not really known; a recent estimate is 5.8 million users in the United States over 18 years of age (**http://www.ora.com/gnn/bus/ora/survey/index.html**). A key sign of the growth and acceptance of the Internet is the URLs now included in advertisements in the mass media, in particular, television and magazines. The Internet is changing so rapidly that the only safe prediction is that the Internet of today will not be the Internet of tomorrow.

Nonetheless, it is possible to project from trends today what the Internet of tomorrow will be like. One trend is faster access. A large fraction of the country is wired for cable TV, and cable "modems" are available; all that remains for the Internet to come to us over TV cable, at Mbps speeds (106 bits/sec), is a connection between the cable operator's equipment and the Internet. Also, phone companies are upgrading their equipment to digital switching technology to enable high speed Internet access. Another trend is wider access. Many PCs sold today come installed with Windows 95 and a modem. Anyone who buys such a computer can connect to the Internet simply by connecting a modular phone plug from the computer to a phone jack. In effect, anywhere there is phone access there is Internet access. The ultimate in Internet access will occur with wireless communications. The technology for cellular phones is easily extended to digital computer communications, so that, literally, Internet access can be obtained anywhere in the country, from Death Valley to Mount Washington.

Specifications are now being developed to extend the WWW into three dimensions. The specifications are called Virtual Reality Markup Language (VRML). VRML browsers and browser extensions are being developed so that an interactive three-dimensional world can be accessed (**http://www.sdsc.edu/SDSC/Partners/vrml/software/browsers.html**). The third dimension will increase the amount of information that a single screen will hold. A student could enter a virtual laboratory and interact in three dimensions with virtual beakers, chemicals, Bunsen burners and other students, like a chemistry version of the popular game Doom (**http://www.islandnet.com/~ccaird/doom/**).

Another technological advance was the development of the Java programming language (**http://www.javasoft.com/**), which allows programs to be run within a World Wide Web browser. Java allows programs to be transferred through the Internet along with text, images and video. Java is an object-oriented programming language; therefore, it is platform independent. Any type of machine will recognize Java programs. A programmer using the C programming language would have to create a separate version of the program for the Macintosh, Windows and UNIX operating systems. Using Java, the programmer writes only one version of the program code, and does not need to worry about what operating system the program will run on. Additionally, programs created with Java run locally on client computers rather than on the HTTP server computer, thereby distributing the computational load. With Java, chemists can create and access interactive, multimedia content from within a World Wide Web browser.

Companies are also creating browser plugins so that files normally requiring helper applications can be run directly within the browser (**http://home.netscape.com/comprod/products/navigator/version_2.0/plugins/**). For example, an interactive presentation created with Macromedia Director can now be sent across the Internet and run locally within a browser (**http://www.macromedia.com/Tools/Shockwave/**). Other plugins appearing recently include a VRML viewer, audio player and Microsoft Word 6.0 viewer.

The Internet is becoming increasingly commercialized. Since it was originally designed for non-commercial use, many of the original Internet users think that Internet access should be unrestricted; that is, all information should be made available to the public free of charge. However, the number of commercial sites on the Internet now exceeds the number of noncommercial sites, and it is clear that commericialization of the Internet will continue. Although the Internet may eventually become a vast electronic infomercial, the Internet infrastructure--that is, the number and quality of high speed connection lines--has improved as a result of the involvement of the commercial sector.

Many Internet resources now contain commercial messages. Popular "free" information resources contain links to site sponsors' home pages (e.g., **http://home.netnscape.com/home/internet-search.html**), commercial firms offer some free services in hopes of enticing users to pay for others (e.g., **http://www.secapl.com/**), and commercial software is distributed in various forms for free on the Internet in hopes of boosting retail sales (e.g., **http://home.netscape.com/comprod/mirror/index.html**).

Two questions arise about the future of education on the Internet. One question is, "Will the Internet replace in-class lecturing?" It is difficult to see that it will in the near future, for the same reason that books have not replaced lecturing. Chemistry educators deliver content by in-class lecturing, assigned book readings, office hours, and so on. The Internet represents another way to deliver content.

Another question is, "Will the Internet replace textbooks?" Again, it is hard to see that it will in the near future. The traditional argument has been that a computer cannot easily be taken to the beach and read. In fact, given the rapid advances in technology, it will soon be possible to do just that. However, many students still feel more comfortable with a textbook than with a computer. Textbooks appear to be a more permanent and reliable reference tool from which to study.

Because the Internet uses public communication pathways (such as airways) and crosses state and international borders, the federal government, through the Federal Communications Commission, will be involved in the future of the Internet. The regulation of Internet content (e.g., **http://rs9.loc.gov/cgi-bin/query/z?r104:S14JN5-1038:**) is